RAND NATIONAL DEFENSE RESEARCH INSTITUTE

Identifying Acquisition Framing Assumptions Through Structured Deliberation

Mark V. Arena, Lauren A. Mayer

Prepared for the Office of the Secretary of Defense

Contents

Acknowledgments

The authors would like to thank Gary Bliss and Mark Husband of the PARCA office for their invaluable input and guidance during the development of this process. Jack Graser and Blaise Durante were the technical reviewers for this work. Their suggestions greatly helped to clarify and improve the document. We thank them for their very useful suggestions.

Abbreviations

AFSS	Air Force Space Surveillance System
APT	advanced pilot training
COTS	commercial off-the-shelf
DoD	Department of Defense
EELV	Evolved Expendable Launch Vehicle
FA	framing assumption
FMS	foreign military sales
FoS	family of systems
GOTS	government off-the-shelf
GPS	Global Positioning System
HMMWV	High Mobility Multi-Wheeled Vehicle
ILS	Instrument Landing System
JPALS	Joint Precision Approach and Landing System
JLTV	Joint Lightweight Tactical Vehicle
JV	joint venture
LCS	littoral combat ship
LEO	low earth orbit
MDAP	major defense acquisition program
MEO	middle earth orbit
OPN	other procurement, Navy

OSD	Office of the Secretary of Defense
PARCA	Performance Assessment and Root Cause Analyses
PEO	Program Executive Office
PM	program manager
RDA	research, development, and acquisition
SBIRS	Space Based Infrared System
SCN	shipbuilding and conversion, Navy
SSA	space situational awareness
SYSCOMS	Systems Command
TBD	to be determined
TSPR	Total Systems Performance Responsibility
VHF	very high frequency

CHAPTER ONE

Structured Process Guidance

Overview

Employment of acquisition framing assumptions (FAs) is an approach to defining and tracking key program assumptions that are made early in program development and throughout the program life. FAs serve as a form of risk analysis to identify uncertainties that may or may not be recognized as such. This document provides supporting information for the briefing titled "Identifying and Prioritizing Acquisition Framing Assumptions: Structured Deliberation Exercise"—a briefing to be used in a group setting to identify FAs. It also provides a brief introduction to the concept of FAs; an overview of the approach used in the briefing to identify FAs; and a discussion of how to operationalize this approach, including an overview of structured deliberation, some of the important concepts in having a successful session, and tailoring questions on program risk areas to help elucidate FAs. Note that the briefing is meant as a starting point and should be tailored to specific program circumstances.

Introduction to Framing Assumptions

In 2013, a RAND report included a definition of acquisition FAs along with a series of program examples.[1] This concept grew out of root cause analysis work by the Performance Assessment and Root Cause Analyses (PARCA) office on Nunn-McCurdy breaches on programs. Many of these breaches were associated with an incorrect or failing foundational assumptions—FAs. By making such FAs more explicit early in the program life cycle and tracking them, the Office of the Secretary of Defense (OSD), and the Services may be able to better manage major risks to and expectations of programs.

An FA is any explicit or implicit assumption that is central in shaping cost, schedule, or performance expectations. FAs may change over the course of execution or new

---

[1] Arena, Doll, and McKernan, 2013.

ones can be added. The PARCA office updated the criteria for FAs[2] as follows (italics in original):

- *"Critical: Significantly affects program expectations.* This criterion means that FAs, when they fail or are incorrect, will have significant cost, schedule and/or performance effects on the program. One possible consequence is a formal program breach. The criterion is meant to exclude the many smaller assumptions made for a program that do not result in significant consequences.
- *No workarounds: Consequences cannot be easily mitigated.* This criterion implies that valid FAs have no workarounds or potential fixes if they are wrong. The consequences of a failed FA will occur. When the FA is wrong, there will be significant cost and/or schedule implications.
- *Foundational: Not derivative of other assumptions.* This criterion is, perhaps, the hardest to understand and define. An FA is foundational if it is a high-level and encompassing assumption. An FA might have derivative assumptions associated with it, but a proper FA will not be derivative or subordinate to other major assumptions. The relationship between foundational and derivative assumptions can be exemplified by the F-35 program.[3] There were at least two important assumptions related to the program that later turned out to be incorrect. The first was that a high degree of concurrency between engineering and production was acceptable. The other was that testing would be more efficient than seen historically. However, these are both derivative assumptions of the true FAs. The foundational FAs were that the design technology was mature and that the competitive prototype was production representative.
- *Program specific: Not generally applicable to all programs.* This criterion implies that FAs should reflect some unique aspects of the program. For example, an FA is not, 'The contract will perform well.' However, an FA might be, 'The key technologies are sufficiently mature such that no component development or prototyping is necessary.'"

In the root cause research, we observed that FAs can be grouped into four major areas, shown in Table 1.1.

Several examples were identified of specific FAs that failed and contributed to Nunn-McCurdy breaches:

- Joint Strike Fighter (F-35)
  - Competitive prototypes are production representative (discussed above).

---

[2] Performance Assessment and Root Cause Analysis Office, 2013.

[3] Blickstein et al., 2011.

Table 1.1
FA Areas

Technological (Component/System Integration)	Management/ Program Structures	Mission Requirements	Cost and Schedule Expectations
Manufacturing expectations	Dependencies on other programs or development efforts	Stability of operational needs	Industrial base/market expectations
Testing expectations		Quantity	Acquisition initiatives or targets
	Contractual incentive strategy/relationships	Capabilities	
Technical approach		Joint needs	
Risk expectations	Organizational management structure	Possibility of a substitute system	Unknown or undefined areas of scope (e.g., facilities locations, support approaches)
Use of simulation		Understanding of threat levels	
Scale of integration	Legal, diplomatic, or political issues		Experience of industry to execute
COTS/GOTS suitability for application	Degree of "Jointness"	Flexibility based on changing intelligence	
Reusability of legacy equipment or subsystems	FMS possibilities		
Technical maturity of components, system, and software			

- The commonality between variants is high and will reduce development and production cost.
- Evolved Expendable Launch Vehicle (EELV)
  - The commercial launch marketplace is robust and can be leveraged for savings.
  - Government launch systems are flexible.
- Space Based Infrared System (SBIRS) High
  - Total Systems Performance Responsibility (TSPR) approach, managed by the contractor, is more effective than a traditional government-led management approach.
  - Software can be reused across development increments.
  - A form of incremental funding will yield large cost savings.

It is important to note that FAs are not necessarily a "bad" thing or incorrect. All programs make assumptions and many may turn out to be valid. The important point is that program outcomes are highly dependent on making the correct FAs. Also, one could go overboard in identifying framing assumptions. The concept is to develop a few concrete, high-level assumptions that can be easily tracked; on the order of three to five assumptions are a good amount for most programs.

In this section, we reproduce the slides from the presentation.

Identifying and Prioritizing Acquisition Framing Assumptions: Structured Deliberation Exercise

July 2014

Group Deliberation Goals

- **Objective**
  - **Identify and prioritize a set of major program assumptions**

- **Three phases:**

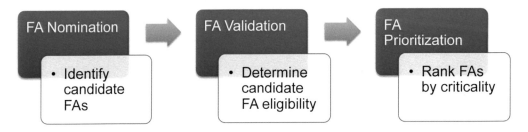

The Group Deliberation Exercise

- Small interactive group exercise to stimulate creative idea generation

- Participants should:
  - Voice/record as many ideas as possible (regardless of their feasibility or unusual nature)
  - Withhold criticism or initial evaluation of ideas
  - Extend, add onto, combine and build upon other participants' ideas

- Some methods are better than others
  - Unrestricted group deliberation has been shown to be less effective than structured, written idea sharing

Sources: Paulus and Yang, 2000; DeRosa, Smith, and Hantula, 2007.

Ground Rules for a Deliberation Session

- Follow a semi-structured, systematic process
- Appoint a facilitator or group leader who does not have a stake in the outcome or strong predispositions; the facilitator should NOT be the program manager
- Gather a small, heterogeneous group
- Allow enough time in the session to properly explore all ideas
- Never immediately dismiss a new idea/minority viewpoint – understand its basis and potential consequences
- Keep an open mind and be willing to change initial opinions
- Use methods that allow all group members to state their opinion such as the sharing of individually written ideas

Sources: Straus, Parker, and Bruce, 2011; Central Intelligence Agency, 2009; DeRosa, Smith, and Hantula, 2007.

Five Steps in the Deliberation Process

- Step 1: "Homework": Read Informational Material (required), identify 3+ candidate FAs (optional)

- Step 2: Overview/Review of FAs and their importance

- Step 3: Identify candidate FAs (*Nomination*)

- Step 4: Determine eligibility (*Validation*)

- Step 5: Rank by criticality (*Prioritization*); identify metrics to track

RAND

Step 1: "Homework"
Read FA Info Paper and Program Background
(required); Identify 3+ Candidate FAs (optional)

1. Read FA Info Paper and FA questions (complete set in back-up slides) prior to meeting for deliberation
2. Read program background
3. Group members come prepared to meeting with 3+ candidate FAs (optional)

RAND

Step 2: Overview of Framing Assumptions and Their Importance

① Definition of an FA and common FA themes
② Importance of FAs
③ Program example FAs

RAND

Definition of a "Framing Assumption"

- *Framing assumption definition: Any explicit or implicit assumptions that are central in shaping cost, schedule, and/or performance expectations*
- *Attributes of a framing assumption:*
  - Critical: Assumption significantly affects program expectations (e.g., cost, schedule, performance).
  - No workarounds: The consequences of an incorrect FA cannot be easily mitigated. The effects are generally outside the project team's control.
  - Foundational: Not subordinate, derivative or linked to other assumptions. The FA may be composed of secondary assumptions, but it is independent from other major assumptions. It represents some central feature of the program.
  - Program specific: Not generically applicable to all programs.
- *Only a few, key framing assumptions per program*

A Framing Assumption is...	A Framing Assumption is not...
A unique aspect of contracting strategy (e.g., competitive prototyping)	The contractor will perform well
Weapon system to be replaced will last until a specified time	Program characteristic (e.g., family of systems)
Use of COTS/GOTS subsystems will save money	Program is affordable

Framing assumptions are not necessarily incorrect or "bad." However, they fundamentally shape program success – a bad assumption can result in significant challenges.

Common Framing Assumption Themes

Technological (Component / System Integration)	Management/ Program Structures	Mission Requirements	Cost and Schedule Expectations

• Manufacturing expectations
• Testing expectations
• Technical approach
• Risk expectations
• Use of simulation
• Scale of integration
• COTS/GOTS suitability for application
• Reusability of legacy equipment or subsystems
• Technical maturity of components, system, and software
	• Dependencies on other programs or development efforts
	• Contractual incentive strategy/ relationships
	• Organizational management structure
	• Legal, diplomatic, or political issues
	• Degree of "Jointness"
	• FMS possibilities
		• Stability of operational needs
		  • Quantity
		  • Capabilities
		  • Joint needs
		• Possibility of a substitute system
		• Understanding of threat levels
		• Flexibility based on changing intelligence
			• Industrial base/ market expectations
			• Acquisition initiatives or targets
			• Unknown or undefined areas of scope (e.g., facilities locations, support approaches)
			• Experience of industry to execute

RAND

Importance of FAs: Failed FAs Have Been Major Contributors to Nunn-McCurdy Breaches

- Joint Strike Fighter (F-35)
  - Competitive prototypes are production representative
  - The commonality between variants is high and will reduce development and production cost

- EELV
  - The commercial launch marketplace is robust and can be leveraged for savings

- Space Based Infrared System (SBIRS) High
  - Total Systems Performance Responsibility (TSPR) approach, managed by the contractor, is more effective than traditional government-led approach
  - Software can be reused between increments
  - A form of incremental funding will yield large cost savings

RAND

Advanced Pilot Training (APT)
Family of Systems (FoS)

Service: Air Force

Program Type: Pre-MDAP

Commodity Type: Aircraft and Ground Training System

Description: Expected to supply an advanced trainer, known as the T-X aircraft, for the fighter/bomber APT track as soon as 2017

Technological	Management/ Program Structures	Mission Requirements	Cost and Schedule
• Training aircraft will be non-developmental (leverage commercial/foreign aircraft's capabilities) with minimal changes.	• Possible synergies with U.S. Navy and foreign militaries. (requirements align and minimal modifications required).	• Use of existing T-38 can be extended until 2020.	• Simulators can be used instead to reduce training flight time (to save money).

RAND

Joint Lightweight Tactical Vehicle (JLTV)

Service: DoD (Army and Marine Corp)

Program Type: Pre-MDAP

Commodity Type: Ground Combat

Description: Will be a successor to the 11 different versions of the High Mobility, Multi-Wheeled Vehicle (HMMWV) that have been in service since 1985

Technological	Management/ Program Structures	Mission Requirements	Cost and Schedule
• Open architecture approach will reduce risk and allow for more efficient upgrades.	• Joint Army and Marine Corps program will save money and requirements are compatible.	• Have effectively assessed long-term vs. short-term needs.	• Competitive prototyping will reduce risk/cost.

RAND

Joint Precision Approach and Landing System (JPALS)

Service: Navy (with Air Force and Army support)

Program Type: MDAP

Commodity Type: Command and Control (Other)

Description: Global Positioning System (GPS)-based precision approach and landing system

Technological	Management/ Program Structures	Mission Requirements	Cost and Schedule
• Modeling and simulation can be used to explore design tradeoffs.	• Navy is best service to lead acquisition due to their more demanding requirements.	• GPS constellation will be robust. • System is suitable for all types of air vehicles. • FAA will move to a GPS-based ILS.	• Ability to leverage COTS/GOTS hardware and software will lower cost.

RAND

Step 3: Identify Candidate Framing Assumptions

① Group members identify and individually record candidate FAs

② Facilitator lists candidate FAs in *FA Matrix*; group members present their FAs to the group

③ Facilitator leads limited discussion

① *Group Members Identify and Individually Record Candidate FAs*

- Group members read seed questions – questions meant to raise issues that may lead to candidate FAs
- Members identify and record candidate FAs, along with a brief supporting explanation for each (these may be prepared ahead of time if Step 1 is implemented fully)
- The seed questions are drawn from each common FA theme; also include more general questions to help identify assumptions; examples include (see back-up slides for complete set):
  - Technical: Have the technologies/types of software employed been used in a similar application, scale, or environment?
  - Management: Is the program's progress dependent on the progress of other programs? Is a new program management approach being adopted?
  - Requirements: Are the requirements stable and well defined?
  - Cost/schedule: Are there any significant savings initiatives/targets assumed?

RAND

② *Facilitator Lists Candidate FAs in FA Matrix; Members Present Their FAs to the Group*

- Facilitator compiles all candidate FAs and lists in the first column of the FA Matrix, eliminating redundant FAs
- Members briefly present their candidate FAs to the group

The FA Matrix

Candidate FA	Eligibility			Prioritize Criticality	Signposts/ metrics to monitor FA
	Program-specific?	No work-arounds?	Foundational?		
FA 1					
FA 2					
⋮					
⋮					
FA *n*					

RAND

③ *Facilitator Leads Limited Discussion*

- **Facilitator leads a limited discussion to clarify or add additional candidate FAs**
- **Facilitator should ensure that all common FA themes are considered when compiling list of candidate FAs**
  - **Any missing FA themes should be revisited**
- **At this time group members*:**
  - **May ask for clarification and rewording of candidate FAs**
  - **Add any new candidate FAs that are uncovered during discussion**

*Group members should not discuss whether the candidate FA is *foundational, program-specific, critical,* or has *no workarounds*

Step 4: Determine Candidate FA Eligibility

① Are candidate FAs program-specific?
② Do candidate FAs have any workarounds?
③ Are candidate FAs foundational?

① *Are Candidate FAs Program-Specific?*

- Facilitator leads a discussion about whether each candidate FA is program-specific?
  - The assumption is not generally applicable to all programs
  - The assumption is not a fundamental physical law or property - such as the gravitational constant.
- Group comes to consensus on whether each candidate FA is program-specific
- If candidate FA is determined to not be program-specific, it is eliminated from eligibility

Candidate FA	Eligibility			Prioritize Criticality	Signposts/ metrics to monitor FA
	Program-specific?	No work-arounds?	Foundational?		
FA 1	Yes				
FA 2	No				
:	Yes				
:	Yes				
FA *n*	Yes				

RAND

② *Do Candidate FAs Have Any Workarounds?*

- Facilitator leads discussion about whether each candidate FA has any workarounds; questions include:
    - Could any of the consequences be avoided through some other means?
    - Could the success of the program still be salvaged through some mitigating action?
- Group comes to consensus on whether each candidate FA has any workarounds
- If candidate FA is determined to have workarounds, it is eliminated from eligibility

Candidate FA	Eligibility			Prioritize Criticality	Signposts/ metrics to monitor FA
	Program-specific?	No work-arounds?	Foundational?		
FA 1	Yes	Yes			
FA 2	No				
:	Yes	Yes			
:	Yes	Yes			
FA *n*	Yes	No			

③ *Are Candidate FAs Foundational?*

- Record all remaining candidate FAs on self-adhesive notes
- With minimal talking, move candidate FAs around into groupings
    - Groupings should reflect some aspect of similarity between FAs
    - Candidate FAs may be moved more than once by different individuals
    - If there is too much back-and-forth on a candidate FA, create a copy of it and place it in two categories
- For each grouping and remaining lone candidate FAs, identify other relevant candidate FAs:
    - Are there related assumptions that have not been identified yet?
    - Is there a new candidate FA that is central to all others in the grouping?
- Identify the foundational FA:
    - Are there similarities within a group?
    - Are some candidate FAs subordinate or a derivative of other FAs?
    - Is the candidate FA or grouping contingent on other events?
    - Does the candidate FA or grouping represent one or more aspects of a more general assumption?

    Source: Institute for Healthcare Improvement, 2004.

Green notes: candidate FAs
Blue notes: foundational FAs

RAND

③ *Example Foundational and Subordinate Candidate FAs*

Littoral Combat Ship Modules (Navy mission systems program developing modules to execute a variety of missions)

- **Candidate FAs that may be grouped together:**
  - **Incremental development will allow new capabilities to be added easily**
  - **Modules are able to be tested on other ship platforms**
- **Foundational FA: Mission modules can be developed independently of sea frames**

RAND

FA Matrix with All Eligibility Items

Candidate FA	Eligibility			Prioritize Criticality	Signposts/ metrics to monitor FA
	Program-specific?	No work-arounds?	Foundational?		
FA 1	Yes	Yes	No → FA 3		
FA 2	No				
FA 3	Yes	Yes	Yes		
FA 4	Yes	Yes	No		
FA 5	Yes	No			
FA 6	Yes	Yes	No → FA 3		
FA 7	Yes	Yes	No → New FA 11		
FA 8	Yes	Yes	No → New FA 11		
FA 9	Yes	Yes	Yes		
FA 10	Yes	Yes	Yes		
New FA 11	Yes	Yes	Yes		

RAND

Step 5: Prioritize Potential FAs by Criticality; Identify Metrics to Track

① Determine criticality of potential FAs
② Determine signposts/metrics to monitor candidate FAs

② *Determine Signposts/Metrics to Monitor Potential FAs*

- **Facilitator leads discussion about signposts and metrics that could be used by the program to monitor the assumption; questions include:**
  - **What events or trends (i.e., signposts) would be expected to occur if this assumption was correct or incorrect?**
    - **How could these events/trends be measured and monitored?**
  - **What metrics could be used to track the validity of the FA?**
    - **Are there threshold values for these metrics that could signify a change in the assumption?**
- **Facilitator records ideas on the *FA Matrix***

Source: Central Intelligence Agency, 2009.

RAND

② *Determine Signposts/Metrics to Monitor Potential FAs*

- **Facilitator leads discussion about signposts and metrics that could be used by the program to monitor the assumption; questions include:**
  - **What events or trends (i.e., signposts) would be expected to occur if this assumption was correct or incorrect?**
    - **How could these events/trends be measured and monitored?**
  - **What metrics could be used to track the validity of the FA?**
    - **Are there threshold values for these metrics that could signify a change in the assumption?**
- **Facilitator records ideas on the *FA Matrix***

Source: Central Intelligence Agency, 2009.

RAND

An Example of a Completed FA Matrix

Candidate FA	Eligibility			Prioritize Criticality	Signposts/ metrics to monitor FA
	Program-specific?	No work-arounds?	Foundational?		
FA 1	Yes	Yes	No → FA 3		
FA 2	No				
FA 3	Yes	Yes	Yes	2	Metric 1
FA 4	Yes	Yes	No		
FA 5	Yes	No			
FA 6	Yes	Yes	No → FA 3		
FA 7	Yes	Yes	No → New FA 11		
FA 8	Yes	Yes	No → New FA 11		
FA 9	Yes	Yes	Yes	3	Signpost a, b
FA 10	Yes	Yes	Yes	4	Signpost c
New FA 11	Yes	Yes	Yes	1	Metric 2, 3

RAND

FA Matrix Template

Candidate FA	Eligibility			Prioritize Criticality	Signposts/ metrics to monitor FA
	Program-specific?	No work-arounds?	Foundational?		

RAND

References

Mark V. Arena, Irv Blickstein, Abby Doll, Jeffrey A. Drezner, Jennifer Kavanagh, Daniel F. McCaffrey, Megan McKernan, Charles Panagiotis Nemfakos, Jerry M. Sollinger, Daniel Tremblay, Carolyn Wong, *Management Perspectives Pertaining to Root Cause Analyses of Nunn-McCurdy Breaches. Vol. 4: Program Manager Tenure, Oversight of Acquisition Category II Programs, and Framing Assumptions*, RAND Corp., MG-1171/4-OSD, 2013.

Irv Blickstein, Michael Boito, Jeffrey A. Drezner, J.A. Dryden, Kenneth P. Horn, James G. Kallimani, Martin C. Libicki, Megan McKernan, Roger C. Molander, Charles Panagiotis Nemfakos, Chad Ohlandt, Caroline A. Reilly, Rena Rudavsky, Evan Saltzman, Jerry M. Sollinger, Katharine Watkins Webb, Carolyn Wong, *Root Cause Analyses of Nunn-McCurdy Breaches. Volume 1, Zumwalt-Class Destroyer, Joint Strike Fighter, Longbow Apache, and Wideband Global Satellite.* RAND Corp., MG-1171/1-OSD, 2011.

Central Intelligence Agency, United States Government, A tradecraft primer: Structured analytical techniques for improving intelligence analysis, March 2009. Retrieved from: https://www.cia.gov/library/center-for-the-study-of-intelligence/csi-publications/books-and-monographs/Tradecraft%20Primer-apr09.pdf

DeRosa, D. M., Smith, C. L., & Hantula, D. A. (2007). "The medium matters: Mining the long-promised merit of group interaction in creative idea generation tasks in a meta-analysis of the electronic group brainstorming literature," *Computers in Human Behavior, 23*(3), 1549-1581.

(IHI) Institute for Healthcare Improvement, (2004), *Idea Generation Tools: Brainstorming, Affinity Grouping, and Multivoting*, Boston, MA. http://www.ihi.org/knowledge/Pages/Tools/BrainstormingAffinityGroupingandMultivoting.aspx

Kebbell, M. R., Muller, D. A., & Martin, K. (2010). *Understanding and managing bias. Dealing with uncertainties in policing serious crime, 16*(1), 87.

PARCA, *Information Paper On Framing Assumptions*, September 13, 20113.

Paulus, P. B., & Yang, H. C. (2000). "Idea generation in groups: A basis for creativity in organizations," *Organizational behavior and human decision processes, 82*(1), 76-87.

Straus, S. G., Parker, A. M., & Bruce, J. B. (2011). The group matters: A review of processes and outcomes in intelligence analysis. *Group Dynamics: Theory, Research, and Practice, 15*(2), 128.

RAND

Seed Questions

The next set of slides include the following seed questions

1. **Topical questions to help identify assumptions**
- **Following common themes: Technical, Management, Requirements, Schedule/Cost**
- **Answering yes to the question might indicate a possible FA for the program in that area. The questions should be tailored to the program**

2. **General questions to help identify assumption**

3. **Questions to assess criticality**

Technical Discussion Questions

- Have the technologies employed been used in a similar application or environment?
    - Is there commercial technology that is being used for the first time in a military application? Who has the data rights?
    - Has the technology demonstrated successfully under the same operating conditions?
    - Has the reliability been demonstrated?

- Does the system depend on COTS solutions or other commercial technologies and services?
    - Is this a novel integration of standard systems?
    - Will these systems require modification for environment (e.g., shock, vibration, electromagnetic, and temperature)?
    - How long might the manufacturer support such an item? Will these services be available over the life of the system? Can the design adapt to component changes/upgrades?

- Is the commercial availability stable? Have all the technologies been demonstrated or successfully operated at the scale planned (e.g,. power density, number of sensors, bandwidth)?
    - How large a scale-up is planned?
    - Are there integration issues at this scale?

RAND

Technical Questions (cont.)

- Are there multiple systems/family of systems (FoS) that need to be integrated?
  - Have similar FoS been successfully integrated?

- Are there new manufacturing methods or techniques involved?

- Are there new or unusual materials involved?
  - Is the source of supply and price stable?

- Have prototypes been developed (or are planned) at the subcomponent or system level?
  - Do the prototypes represent something close to a production configuration?
  - Has the prototype effort resulted in reductions to cost and/or schedule?
  - Have the prototypes demonstrated needed technical maturity?

- Is the software development well understood?
  - What is the size of the overall programming effort?
  - How many lines of code per day can we expect? What error rate?
  - What are the assumptions around software reuse?

Management Questions

- Novel management structures
  - Is the government acting as system integrator?
  - Are multiple PEOs/PMs involved?
  - Do industry partners participate through new commercial partnerships or JVs?

- Is the program's progress dependent on the progress of other programs?

- Are there unique legal, diplomatic, or security issues?

- Does the program have an experienced workforce? Will there be issues retaining this workforce?

RAND

Requirements Questions

- Is there joint/foreign involvement?
  - Are the program requirements compatible between the stakeholders?
  - Does each participant require a customized version?
  - Is there uncertainty with respect to quantities for partners?

- Are the requirements stable and well defined?

- Will capability be met through an evolving design or series of upgrades?

- Is the need well understood (both from a capability sense and timing)?

- Are there unknown major areas of scope, e.g., facilities locations, operational availability, support equipment/infrastructure?

- Could another system substitute for this one?

- Do all the requirements need to be addressed for the program to be successful?

RAND

Cost and Schedule Questions

- Does the program rely on sole source(s)?

- Have the intellectual property rights been resolved?

- Are there workforce supply or demand issues? For instance, will the program contractor or vendor require significant hiring? Are key workforce skills/ trades in short supply? Can we hire at the rate based in our plans for that location?

- Is the stability of the vendors and suppliers base understood? Are there key suppliers who are at risk?

- Has the prime contractor executed a similar program (either in complexity or system/commodity type) before?

- Is there going to be a management reserve/should there be cost targets for the program?

- Has sufficient time been allowed to get the necessary approvals from OSD?

Cost and Schedule Questions (cont.)

- In the case of COTS products/solutions – is the commercial marketplace stable in terms of demand and pricing or is it cyclical? Can you obtain the data needed from the COTS supplier?

- Are there any significant savings initiatives/targets assumed?
  - What is the source of the savings?
  - Are they reasonable based on experience on existing or completed programs?

- Are there schedule pressures or tight deadlines to meet an IOC date?

- Is the testing plan adequate?
  - Is the time allotted comparable to previous efforts?
  - Is the number of test articles similar?
  - Is there a reliance on simulation to supplant some of the testing?
  - Does the test community have the capacity in the timeframe needed?
  - Is there particular test equipment needed? Does it need to be developed?

General Questions to Identify Assumptions

Consider the following questions when determining potential framing assumptions from seed questions

- Does information exist that:
    - Disconfirms or is contradictory to this judgment?
    - Was previously dismissed that might now be relevant to the topic?
    - Is new and could change this judgment? Has it been properly adjusted?
- How accurate and reliable is information upon which this judgment is based?
    - Was incomplete, imprecise, or ambiguous information used?
- Could certain circumstances (e.g., social, technological, economic, environmental, political, organizational) affect this judgment?
    - Does the judgment account for these circumstances? How sensitive is it to these circumstances?
    - Could circumstances proceed differently than expected?
    - For what circumstances would this judgment be abandoned?
    - Have all plausible but unpredictable circumstances been considered?

Sources: Straus, Parker, and Bruce, 2011; CIA 2009; Kebell, Muller, and Martin, 2010.

Criticality Questions

- If this assumption was incorrect:
  - To what degree would it affect the success of the program?
    - Could it delay the program schedule?
    - Could it change program requirements?
    - Could it affect program performance?
    - Could it directly increase program costs?
  - Could this result in other/multiple consequences to the program?

- Would changes in this assumption result in the contradiction of other identified assumptions?

Source: Central Intelligence Agency, 2009.

Further Examples

Littoral Combat Ship (LCS) Mission Modules

- *Service:* Navy

- *Program Type:* Pre-MDAP

- *Commodity Type:* Mission Systems

- *Description:*

  - Provides Combatant Commanders assured access against littoral threats

  - Mission systems are added to the mission module baseline incrementally as they reach a level of maturity necessary for fielding

  - Uses evolutionary acquisition process

  - Provides an open architecture environment that enables future rapid insertion of new technologies

- *Last Milestone Awarded:* Milestone A (May 27, 2004)

Sources: DAES (as of April 17, 2012) (FOUO); "LCS Mission Modules: Training Strategy Increasing Modularity for Maximum Adaptability Brief for Implementation Fest 2010" (August 10, 2010)

RAND

Littoral Combat Ship (LCS) Mission Modules

Technological	Management/ Program Structures	Mission Requirements	Cost and Schedule
• Independent development of sea frames and modules • Spiral/ incremental development will lower risk. • Ability to successfully test modules on other ship platforms. • New capabilities can be added easily.	• New business model • Benefits of open architecture/ commercial practices. • Government suited to act as system integrator. • RDA can be program focal point (4 PMs, 3 PEOs, 2 SYSCOMS).	• Willingness of Navy to drop requirements (in spiral context) to keep to schedule.	• Modules can be funded OPN instead of SCN.

RAND

Space Fence

- *Service:* Air Force

- *Program Type:* Pre-MDAP

- *Commodity Type:* Radar

- *Description:*

    - Provides a radar system operating in the S-band frequency band to replace the AFSSS VHF "Fence" radar that currently performs detection of orbiting space objects.

    - The S-band radar will have a modern, net-centric architecture that is capable of detecting much smaller objects in LEO/MEO.

    - The system will operate with greater accuracy and timeliness to meet warfighter requirements for SSA.

    - Two radar sites are planned, with locations TBD.

- *Last Milestone Awarded:* Milestone A (March 14, 2009)

    Sources: DAES (as of April 17, 2012) (FOUO); Mar 14, 2009 ADM (FOUO)

RAND

Space Fence

Technological	Management/ Program Structures	Mission Requirements	Cost and Schedule
• Capability is achievable despite some immature technologies at outset. • Can scale technology to track order of magnitude greater number of objects (radar components, software interoperability.	• Ease and implications of obtaining host nation agreement for siting. • Block approach a more effective acquisition strategy.	• Minimal manpower required to operate and support system. • Two of the original three sites will be sufficient.	• Competitive prototyping reduces risk and cost.

RAND

References

Arena, Mark V., Abby Doll, and Megan P. McKernan, *Exploring the Concept of Framing Assumptions for Major Acquisition Programs,* Santa Monica, Calif.: The RAND Corporation, unpublished research.

Arena, Mark V, Irv Blickstein, Abby Doll, Jeffrey A. Drezner, Jennifer Kavanagh, Daniel F. McCaffrey, Megan McKernan, Charles Panagiotis Nemfakos, Jerry M. Sollinger, Daniel Tremblay, and Carolyn Wong, *Management Perspectives Pertaining to Root Cause Analyses of Nunn-McCurdy Breaches. Volume 4: Program Manager Tenure, Oversight of Acquisition Category II Programs, and Framing Assumptions,* Santa Monica, Calif.: The RAND Corporation, MG-1171/4-OSD, 2013. As of September 19, 2014:
http://www.rand.org/pubs/monographs/MG1171z4.html

Blickstein, Irv, Michael Boito, Jeffrey A. Drezner, James Dryden, Kenneth Horn, James G. Kallimani, Martin C. Libicki, Megan McKernan, Roger C. Molander, Charles Nemfakos, Chad J. R. Ohlandt, Caroline Reilly, Rena Rudavsky, Jerry M. Sollinger, Katharine Watkins Webb, and Carolyn Wong, *Root Cause Analyses of Nunn-McCurdy Breaches. Volume 1: Zumwalt-Class Destroyer, Joint Strike Fighter, Longbow Apache, and Wideband Global Satellite,* Santa Monica, Calif.: RAND Corporation, MG-1171/1-OSD, 2011. As of September 19, 2014:
http://www.rand.org/pubs/monographs/MG1171z1.html

Central Intelligence Agency, United States Government, *A Tradecraft Primer: Structured Analytic Techniques for Improving Intelligence Analysis,* March 2009. As of September 19, 2014:
https://www.cia.gov/library/center-for-the-study-of-intelligence/csi-publications/books-and-monographs/Tradecraft%20Primer-apr09.pdf

Defense Acquisition Executive Summary, "LCS Mission Modules: Training Strategy Increasing Modularity for Maximum Adaptability Brief for Implementation Fest 2010," August 10, 2010. Not available to the general public.

DeRosa, D. M., C. L. Smith, and D. A. Hantula,"The Medium Matters: Mining the Long-Promised Merit of Group Interaction in Creative Idea Generation Tasks in a Meta-Analysis of the Electronic Group Brainstorming Literature*," Computers in Human Behavior*, Vol. 23, No. 3, 2007, pp. 1549–1581.

Institute for Healthcare Improvement, *Idea Generation Tools: Brainstorming, Affinity Grouping, and Multivoting,* Cambridge, Mass., 2004. As of September 19, 2014:
http://www.ihi.org/knowledge/Pages/Tools/BrainstormingAffinityGroupingandMultivoting.aspx

Kebbell, M. R., D. A. Muller, and K. Martin, "Understanding and Managing Bias," *Dealing with Uncertainties in Policing Serious Crime*, Vol. 16, No. 1, 2010, p. 87.

Paulus, P. B., and H. C. Yang, "Idea Generation in Groups: A Basis for Creativity in Organizations," *Organizational Behavior and Human Decision Processes*, Vol. 82, No. 1, 2000, pp. 76–87.

Performance Assessment and Root Cause Analysis Office, *Information Paper on Framing Assumptions,* Washington, D.C., September 13, 2013.

Straus, S. G., A. M. Parker, and J. B. Bruce, "The Group Matters: A Review of Processes and Outcomes in Intelligence Analysis," *Group Dynamics: Theory, Research, and Practice,* Vol. 15, No. 2, 2011, p. 128.